Table of Contents

Introduction	----------------------	Page 1
The Zoologist	----------------------	Pages 2~5
The Botanist	----------------------	Pages 6~9
The Ecologist	----------------------	Pages 10~13
The Microbiologist	----------------------	Pages 14~17
The Biochemist	----------------------	Pages 18~21
The Chemist	----------------------	Pages 22~25
The Physicist	----------------------	Pages 26~29
The Astronomer	----------------------	Pages 30~33
The Geologist	----------------------	Pages 34~37
The Geographer	----------------------	Pages 38~41

Introduction

What do you want to be when you grow up? This is a question asked to a kid a million times over the ages. Can you still remember your answer? Mine is pretty simple. When I grow up, I want to be a Teacher. My playmates would say a Doctor, a Policeman or an Engineer. As a kiddos, we thought nothing much about the world of grown ups. And that is good. Back then, what's most important is to grow healthy and happy.

As you grow older into teens, you start to know more about the world of grown ups. Terms like jobs, career paths, responsibilities and even salary, how much adults are earning each month. It is not too early to start exploring career opportunities. Teens have to tap on their strengths and interest to start exploring relevant career options. Whether you are someone who loves to create, investigative or collaborative there is always something just right for you.

The main objective for writing this book is for you to have an insight of the different career paths out there. This will be a series, starting with Life and Physical Sciences jobs. Hopefully after reading this book, you will have something to think about as you move into higher education. The early you know what you want to become, the early you can prepare yourself with skills and interest. Don't rush, but enjoy the process. Happy exploring!

The Zoologist

Zoologists work with animals. Yes, all animals whether small or big, feathered or scaly, gentle or dangerous. They study how the animals live, their anatomy, characteristics and behaviours and interactions with other animals or their surrounding.

The animal kingdom is composed of millions species. The Zoologist will never run out of subject to study. The animal kingdom is divided into three categories; insects, vertebrates and mollusks. The five most well known classes of vertebrates are mammals, birds, fish, reptiles and amphibians.

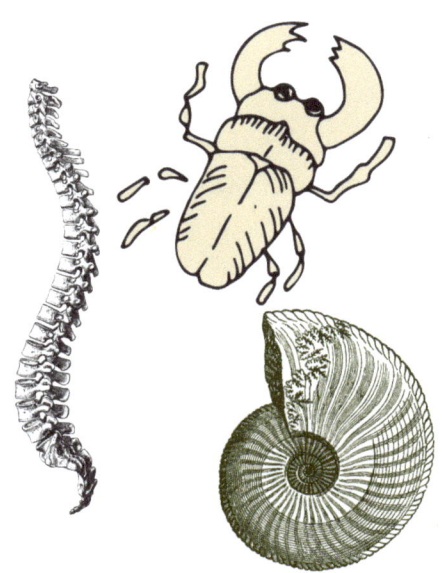

Animals plays a huge part in our ecosystem. Animals are both predator and prey. Balanced ecosystem must have a correct predator to prey ratio.

If there are too many flies than frogs; flies would be everywhere. And we don't like that to happen. Zoologist and Ecologist works together to help the preservation of animal species from extinction and the same time prevent over population.

A Zoologist can work as zookeeper, researcher, educator and more. Zoologist does research about animals and use the data to help understand the animal reproduction and diseases. As we all know they also work in the zoo to train and take care of the animals.

Zookeeper	Researcher

Observation Skill is important to have for a good Zoologist. Animals don't speak a language that we know; Zoologist must spend most of their time observing the animal to understand them.

Dr. Jane Goodall is one of the most famous Zoologist in the 20th century. Considered the world's most expert on chimpanzees. She spent nearly 60 years of her life researching & understanding wild chimpanzees in Tanzania, Africa.

Aristotle (384-322 B.C.) is one of the oldest Zoologist. A Greek Philosopher and a Polymath, a student of Plato. He divided the animals into two groups; those with blood and those without blood. He also has an accurate anatomy of an octopus and cuttlefish. He is the tutor of the famous Alexander the Great.

Zoologist workplace can be a Zoo, a Laboratory, a forest or in the Wild. Some of the most studied and diverse ecosystem in the world is a workplace for a Zoologist. Places like the Amazon Rainforest, the Alaskan Backcountry, the Everglades in Florida, Safari's in Africa and Sundarbans in India and Bangladesh.

The animals are counting on you to protect them. This is not an easy job but a rewarding one. Do you think you have the interest to be a Zoologist?

The Botanist

A Botanist is a plant Scientist who studies the biology of plants and other organisms such as fungi and algae. They study the way plants grow and reproduce. Botanist also studies the effect of the human activities to plants.

It is important to study plants because it is our source of food and medicine, provides the Oxygen that we breath and shelter to both animals and humans. Plants are one of the most important part of our ecosystem. Plants also maintains the climate and lower air pollution.

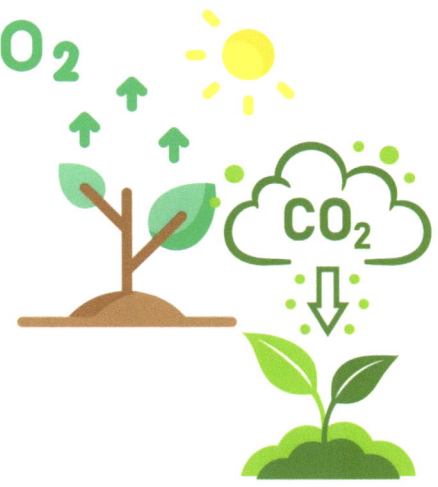

Carl Linnaeus (1707-1778), a Swedish Biologist and Physician is one of the most famous Botanist in history. He is the father of Taxonomy. He developed a system in naming plants and animals called binomial system.

Just imagine for a minute a world without plants. We will all be dead, it's that serious. Botanist plays a big role in our society. Another important use of plant is by helping to reduce the carbon level in the environment and helps reduce the global warming. Trees converts a lot of Carbon to Oxygen, cleaning the air that we breath in. They are our best friends. Plant a tree today for a cleaner and greener future.

Botanist study what makes up a plant so we can use it to produce ingredients for making medicines. Example is a pineapple, which is a source of Bromelain, an enzyme used in medicines for reducing swelling.

They also study how to increase the crop production. Plant can also produce biofuels, not only energy for our body, but energy to fuel machines too.

A Botanist can be specialised in many studies like Taxonomy, Ecology, Forestry or Biology. And there are a long list of studies they can specialised on. Like the zoologist, they will never ran out of interesting things to study.

Biology is the study of interactions between plant cells in relation to the plant's daily functions and reproductive activities. Biology alone branched to a lot of research areas like genetics, molecular biology, biochemistry and more.

Ecology is the study of plants in relation to their environment. One of the most interesting area of research is Marine Ecology, the study of plants underwater. Oceans and deep seas have some of the most beautiful and diverse ecosystem.

Where do Botanists work? Most Botanist will work in Private companies, a non-profit organization or Government agencies. They can work in industries related to Agriculture, Environmental conservation, forestry, pharmaceuticals and food science.

Botanist in the lab studying the plant structure and chemical or medicinal properties. They use lab instruments for biological testing.

Botanist in the field studying plant growth, diseases, population and diversity. The more diverse the ecosystem, the better.

This job is cool, diverse and has high impact in our society. Are you an analytical and collaborative person? Do you have passion in raising awareness about environmental concerns and plant conservation? To succeed as Botanist you should have good research and communication skills. Do you think you have the interest to be a Botanist? Think about it!

 # The Ecologist

An Ecologist is a Scientist who studies the natural ecosystems and the living things that live in them. They analyse data on plants, animals and environmental conditions.

They study the impact of human activities like, construction, farming and manufacturing on our environment particularly lands and waterways.

Ecologists look at the laboratory test results whether a land or water is polluted. They then worked on rehabilitation and future conservation.

Making a clean and safe habitats for all living things is the mission of an Ecologist. Their main role is how to maintain balance and diversity in our ecosystem.

Ecologist works on big environmental problems like acid rain, species extinction, overfishing, overpopulation, pollution, global warming, climate change and more.

Acid Rain is the term used to describe any form of precipitation with acidic components that falls to the ground from the atmosphere. This can be rain, snow, fog or even dust.

This is produced from the emission of sulfur dioxide and nitrogen oxides from human activities mostly from combustion of fossil fuels like in cars and factories. Acid rain contaminates water sources and habitats. Also weakening trees and depletes soil's important nutrients.

The same burning of fossil fuels such as coal, oil and natural gas causes a global warming. Cutting down of trees too.

These are just few of the long list of environmental problems caused by humans through overconsumption, industrialisation and globalisation. The impact have accumulated through years and years of neglect. The farther the product travels, the more fuel is consumed, and greater level of greenhouse gas emissions is produced.

Ecologists worked on solving these environmental issues. But it is our responsibility too. The scale of the problem is so big, it takes decades before we can see a change. We must do our share and help our Ecologists do their job.

What can we do? Remove the cause of the problem like toxic chemical waste. Our share is to only use earth friendly and recyclable products; like less plastics from now on.

Restoring the damaged ecosystems like replanting and reseeding. Our share is to plant a tree today and join community clean up activities.

Rachel Carson (1907~1964) is a famous Ecologist who authored a book called "Silent Spring" to show how humans are damaging the planet we live in. Since we are the cause of the problem, let we be part of the solution also.

Eliminating the effect of globalisation is a dream that probably won't come true. But we still have a chance to minimize the problem. It requires a collective effort of every person in the planet. Our Ecologists will lead the way to a well-thought, unbiased plans for control and rehabilitation.

There are many specialisations in Ecology such as Urban, Industrial, Landscape, Population, Aquatic and more.

Population Ecology studies factors that affects the species population and how it affects the environment. Remember the frog and fly example?

Conservation Ecology is the study of preservation and management of our natural resources like forest, land, water, mine or sea and the biodiversity of our ecosystem.

Aquatic Ecology is the study of ecosystem (plants and animals) found in the water bodies like river, lakes and streams.

As the world becomes global, so is our problem. The job of an Ecologist is not easy but truly rewarding. Working for the betterment of our planet is an honourable job. Do you think you have the interest to be an Ecologist? Think about it!

The Microbiologist

A Microbiologist studies microscopic organisms like bacteria, algae, fungi & parasites. They investigate growth, structure, interactions and other characteristics of these life forms.

Bacteria are microscopic living organism that have only one cell. Some bacteria are not harmful and useful but some are harmful that can make you sick.

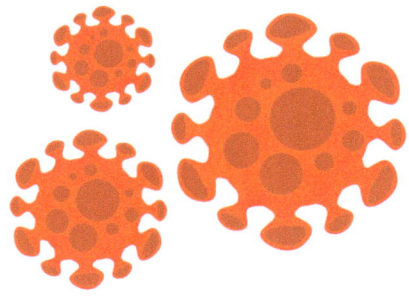

Virus are mostly smaller than bacteria. They are infectious agent that replicates only inside the living cells of an organism.

Fungi are eukaryotic organisms (cells that contains a nucleus) that include microorganisms such as yeasts, moulds and mushrooms.

The study of microbiology includes the following fields. **Bacteriology** is the study of bacteria; it's shapes and arrangements, its environment, it's traits and chemical substances and processes that occurs in the bacterial cell. **Mycology** is the study of fungi. **Protozoology** is the study of protozoa. **Phycology** is the study of algae and **Parasitology** is the study of parasites, their hosts and the relationship within them.

Louis Pasteur (1822–1895): The Master of Microbiology is a French chemist, pharmacist & microbiologist. He is famous for his discoveries of principles of vaccination, actions of enzymes in microbial fermentation & food preservation to eliminate pathogens also known as pasteurisation.

Antoine van Leeuwenhoek (1632–1723): is a Dutch Microbiologist. He is known as the father of Microbiology. He is best known for his works on microscopy. He was the first person to observe and experiment with microbes, using his self design and made microscope.

Ferdinand Julius Cohn (1828–1898). Is a Pioneer of Bacteriology. He is a German-Polish Biologist who first classify algae as plants. He established the use of sterile culture mediums.

Microbiologist conducts microbiology laboratory activities, including sample collection and microbiological testing.

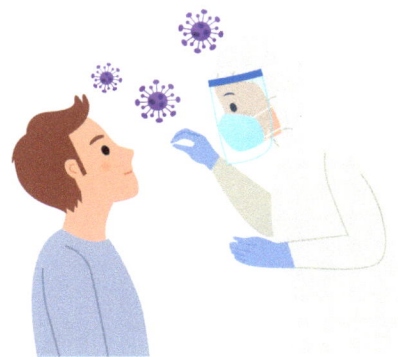

Other responsibilities includes doing systematic documentation, maintain instrument accuracy, perform environmental monitoring of production areas and laboratory.

Microbiologist have opportunities for employment in government, hospitals, public health laboratories, research laboratories, forensic laboratories and industrial laboratories (food, dairy, chemical, pharmaceutical, and genetic engineering companies).

Microbiologist can work as Food Technologist, Infectious disease physician, Lab technician, medical technologist, Microbiology professor, research scientist, water treatment specialist , etc.

Microbiologist requires good knowledge in science and laboratory skills. It is an interesting job which help people in the most important need of humans: food and health. Do you have an ever-inquiring mind? Or perhaps you love problem-solving and doing experiments. This might be the future you. Think about it!

The Biochemist

Biochemists, sometimes called molecular biologists or cellular biologists, are scientist who are trained in biochemistry. The study of chemical processes and transformations in living organisms. They study DNA, proteins and cell parts; such as cells development, growth, reproduction, heredity and disease. DNA are instructions for how to make the body, like recipe for baking cupcakes.

Others study the genetic variation of plants and animals, to understand how genetic traits are carried through successive generations.

Biochemists study the nature of the immune system. How the information encoded in the genes of living organisms instruct the assembly of proteins. And they also study how to isolate, analyze and synthesize different biochemical products and processes. That sounds very scholarly, but don't be taken aback.

Biochemists work in basic and applied research. Basic research is conducted without any immediately known application; the goal is to expand human knowledge. Applied research is directed toward solving a particular problem like the case of Covid-19.

They work to understand how certain chemical reactions happen in tissues and record the effects of medicines, food, allergens and other substances on the living tissue. The aim of a biochemist is to improve our quality of life by understanding living organisms at the molecular level.

Their job can include running laboratory experiments to develop effective medicines or going out in the field to collect cell samples from animals and plants.

Biochemists must also prepare technical reports after collecting, analyzing and summarizing the information and trends found. One must have a good written and oral communication skills to share the technical reports to other scientists or to their teams.

Biochemists are typically employed in the life sciences, where they work in the pharmaceutical or biotechnology industry in a research role. There are different field to focus on such as medicine, agriculture, veterinary science, environmental science, forensic science, manufacturing and more.

Clinical biochemists can work in hospital laboratories to understand and treat diseases, and industrial biochemists can be involved in analytical research work, such as checking the purity of food and drinks.

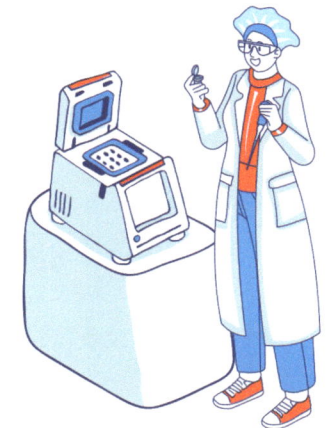

Biochemists in the field of agriculture research the interactions between plants, animals, microorganisms and herbicides. Understanding all these interactions is needed for designing a sustainable farming practice and maximize crop yield.

Forensic biochemist analyse biological information, like DNA or blood samples from the crime scene to help solve the criminal case.

Interesting isn't it? It maybe a little hard, but if you put your heart on it and be diligent in your study, you can do it. Here are some of the skills required for this job. Have a look and ask yourself whether this one is for you.

Some of the job skills and abilities that one needs to attain to be successful in this field of work include science, mathematics, reading comprehension, writing, analytical and critical thinking, problem solving and more. These skills are critical because of the nature of the experimental techniques of the occupation.

One of the most important skills are **Analytical and Critical thinking skills**. Biochemists must be able to conduct experiments and analyses with accuracy and precision. They make conclusions from experiment results through sound reasoning and judgement.

Math skills because they use complex equations and formulas regularly in their work. They need a broad understanding of math, including calculus and statistics. **Problem solving skills** to find solutions to complex scientific problems. **Perseverance is important**. Research works involves lots of trial and error; and they must not be discouraged if they fail. Failures are our greatest teacher. Some of the great discoveries comes after many failures.

The Chemist

Chemistry is one of the oldest branch of Science. It is the scientific study of the properties and behaviour of matter. Chemists conduct experiments in labs in order to analyze substances, develop new products or improve existing ones. Substance are analyzed to determine their chemical composition and concentration of elements for various applications.

The Periodic Table of Elements

The periodic table is the ordered arrangement of all known elements present in the earth. It is the icon of chemistry and is also used by other sciences. The atom is the basic unit of chemistry. The first generally accepted periodic table was that of Russian chemist Dmitri Mendeleev in 1869.

Chemist may specialize in one or more areas, such as organic chemistry, inorganic chemistry, physical, analytical and biochemistry.

Organic chemistry is the study of the structure, properties and reactions of chemical compound that contains carbon-hydrogen or carbon-carbon bonds.

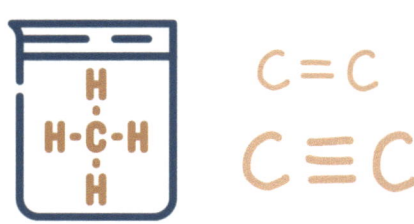

Inorganic chemistry is the study of synthesis and behaviour of inorganic compounds. It is usually found in earth as minerals. It is opposite of organic since it does not have carbon-hydrogen bonds.

Chemistry acts as a foundation for understanding other scientific disciplines. Chemistry explains the plant growth in Botany, the formation of igneous rocks in Geology, how the atmospheric ozone is formed in Ecology, how the medications work in Pharmacology and more. Here are some of the chemical reactions in our everyday life.

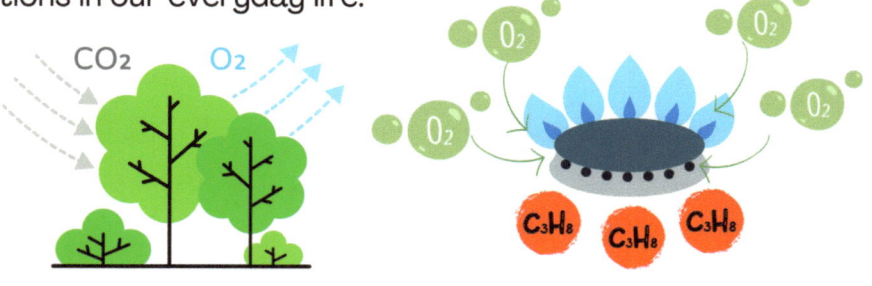

Photosynthesis   Combustion or Burning

A chemical reaction is a transformation of some substances into one or more different substances. During the reaction, the atoms are re-arranged and is accompanied by an energy change as new products are formed.

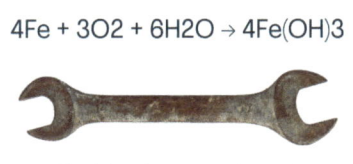

Oxidation or Rusting

Fireworks

Marie Curie (1867–1934) was a Polish and naturalised French Physicist and Chemist. She is famous for her research work on radioactivity. She is the first woman to win the Nobel Prize and the first person to win it twice. She and her husband Pierre Curie discovered the radioactive metals Polonium and Radium.

Antoine-Laurent de Lavoisier (1743-1794) was a French chemist and the "Father of Chemistry." Lavoisier is well known for his discovery of the role of Oxygen to combustion, and his description of the chemical composition of air. He also discovered Silicon.

Here are some roles of Chemists. (1) Analyze chemical substances, (2) Customise chemical formula, (3) Draft reports of the experiments (4) Monitor laboratory equipments to adhere to regulations, (5) Develop and conduct test. The major employers of chemists are academia, pharmaceutical and chemical industries and government laboratories. Below are some of the laboratory equipments a Chemist used in their daily work.

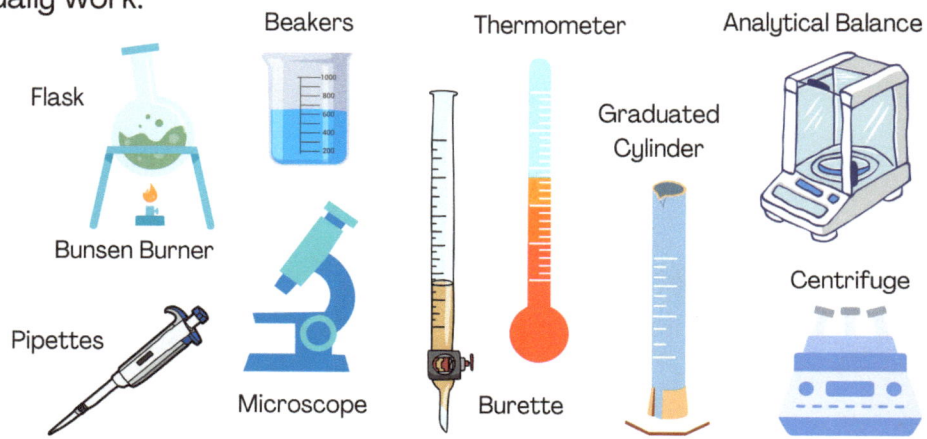

Because chemicals are reactive, a distinct disadvantage of chemistry careers is the risk of exposure to toxic chemicals and biological agents, volatile organic compounds and compressed (high-pressure) gasses. But all these can be overcome by following the safety practice.

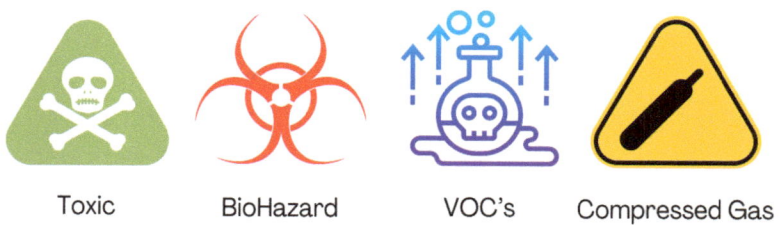

This job is full of actions. and have a tremendous impact in our society. Do you think you have the interest to be a Chemist? Think about it!

The Physicist

Physics is the study of matter, it's motion and behaviour through space and time and other entities like energy and force. Physicists study the behaviour of the physical world and find practical ways to apply new knowledge gained from their research in areas of science and technology. Physicists are usually identified within three broad roles.

Theoretical Physicists develop theories or models of how particular aspects of the world works.

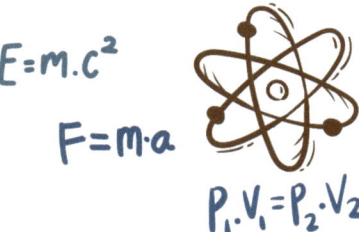

Experimental Physicists test these theories, determining their limits and suggesting new approaches to them.

Gravity   Bouyancy

Applied Physics is the application of Physics to solve scientific or engineering problems. Physicists design, develop and introduce new technology.

Physicists work across a wide range of research fields such as sub-atomic physics, biological physics and cosmological physics. There are twenty over branches of physics to choose their specialisation from.

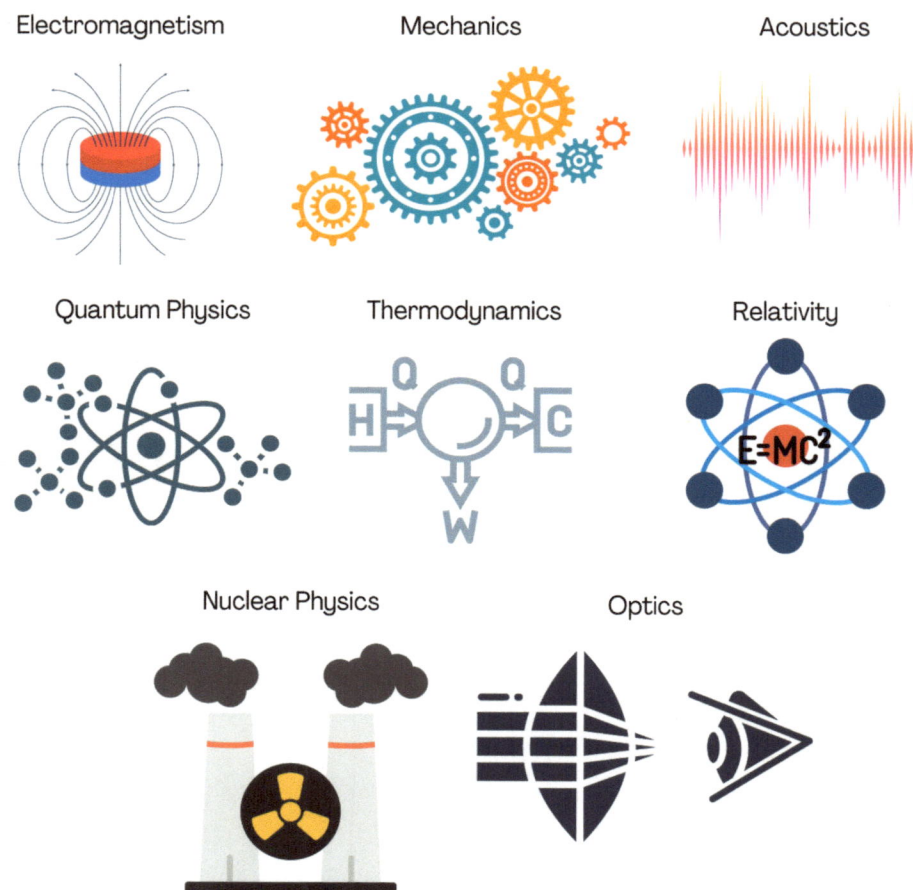

The study and practice of physics is based on the knowledge bank of discoveries and insights from ancient times to the present. Many mathematical and physical wisdom and ideas used today found their earliest origin in the work of earliest civilizations, such as the Babylonian astronomers and Egyptian engineers, the Greek philosophers of science and mathematicians.

The modern physics is from the scientific revolution in Europe. Starting with the works of Copernicus leading to the physics of Galileo and Johannes Kepler.

Isaac Newton (1643-1727) was an English Polymath, active as mathematician, physicist, astronomer, alchemist, theologian and author. He identified the concept of gravity and the theory of mechanics.

Michael Faraday (1791-1867) was an English scientist best known for his work on electricity and magnetism. He made the first electric generator. He established the theory of electromagnetic fields in physics and the concept of magnetism affecting the rays of light

Our list will not be complete without our friend **Albert Einstein** (1879~1955), a key theoretical physicist in the 20th century who developed the theory of relativity and parts of early quantum theory.

Typical duties of physicists includes research, experimentation, observation and data analysis, data preparation and instrumentation. They also do design and development of industrial or medical equipment, and more. The three major employers of career physicists are academic institutions, laboratories, and private industries.

Instrumentation   Design & Development   Data Analysis

Here some of the skills a Physicist must have to succeed. (1) **Logic** and **measurable reasoning** to construct feasible hypotheses. (2) **Curiosity** is the driver which push Physicist to constantly learn and expand their knowledge. (3) One must be good in **Mathematics** because Physicist always deals with complicated calculations to prove their theory. (4) **Problem solving** skills should be topnotch because the scientific profession's role is to find answers to science-related questions.

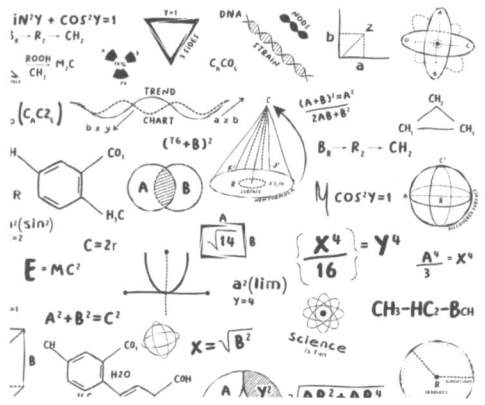

This career seems to be overwhelming for a young teen. But don't drop it yet. You still have few more years in school to assess whether you have the right skills and interest.

The Astronomer

An astronomer is a scientist who studies outer space to find scientific answers to many questions regarding the nature of the universe. They observe celestial objects such as stars, planets, moons, comets and galaxies. Examples of topics or fields astronomers study include planetary science, solar astronomy, the origin of stars, or the formation of galaxies.

There are two main types of astronomers. **Observational** Astronomers make direct observations of celestial objects and analyze the data. While, **Theoretical** Astronomers create and investigate models of things that cannot be observed. They use the observed data to create models or simulations to hypothesize how celestial objects work.

Here are some of the famous Astronomers who pioneered in understanding of the universe we are in. Their work became the foundation of modern Astronomy.

Nicolaus Copernicus (1473-1543) was a Renaissance Polymath who formulated the "Helioncentrism" theory; means the sun is the central point while the earth and other bodies revolves around it. He is known as the Father of Astronomy.

Galileo Galilei (1564-1642) was an Italian astronomer, physicists and Engineer. Considered the founder of modern science. He stirred up and transformed the understanding of the universe. He invented the thermoscope and a hydrostatic balance. He was also the inventor of different military compasses.

Johannes Kepler (1571-1630) was a German astronomer, astrologer, physicist and mathematician who is famous for his three laws of planetary motion and law of continuity. He discovered that the planets move in ellipse around the sun.

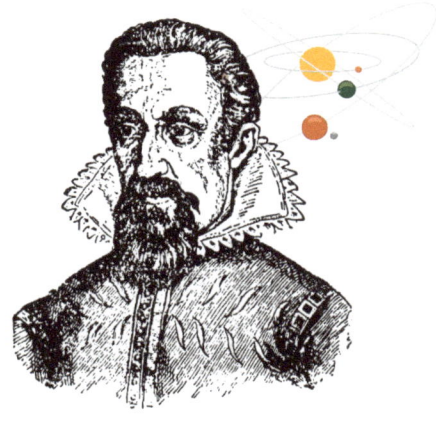

The main equipment an Astronomer need to master is the telescope. Astronomers spend little time observing the celestial bodies. Most of their time are spent on data analysis of what they observed. With modern and advance technology, using high computing systems in mathematics and physics, Astronomers can make predictions faster.

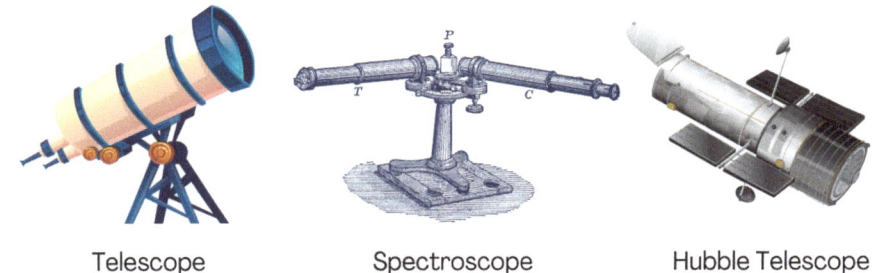

Telescope   Spectroscope   Hubble Telescope

The observatory is where the Astronomers usually work. It usually house advanced telescopes and other astronomical equipments. Astronomical observatory can be ground-based, airborne, underground-based or space-based like the Hubble telescope.

Edwin Hubble was an American Astronomer who discovered that there are other galaxies in the universe beyond the Milky Way. He also observed that the universe is expanding at a constant rate.

Astronomers spend most of their time working on research. They also have other duties like teaching, helping observatory daily operations and building instruments.

Astronomers also studies radio waves, X-rays and cosmic rays. Another area of interest is the study of blackholes and Neutron Stars. Neutrinos is the most abundant particles that have mass in the universe.

Others monitor space debris that could interfere with satellite operations. Many physicists and astronomers work in applied research. They use their knowledge to develop technology or solve problems in areas such as energy storage, electronics, communications, and navigation.

Professional astronomers are highly educated individuals who typically have a PhD in physics or astronomy. One needs to be skilled in Math, Physics & other branch of Science. Like most of science disciplines, an Astronomer must have skills in analytical-thinking, logic and reasoning. Communication skills both written and verbal is also required.

The universe is still expanding as of now when I write this book and by the time you read this. That means our work in understanding the universe is far from over or we can say that it will never end. The universe still offers so many questions, so many discoveries waiting to happen. Could it be waiting for you to discover? Think about it!

The Geologist

Geology is the study of the Earth and everything it is composed of. It looks at how the earth was formed, it's nature, materials and processes. A Geologist is a scientist who studies geology; solid, liquid and gaseous matter that makes-up the earth. They also study earth science and geophysics. There are many types of Geologist.

Environmental Geologist study the human impact on the Earth systems. They are leading the damage prevention and mitigation from natural disasters like earthquakes, volcanoes eruptions, tsunamis and landslides.

Economic Geologist explores and develop earth's resources. They usually work in the energy and mining industry searching for natural resources like petroleum, natural gas and metals.

James Hutton (1726-1797) a Scottish naturalist is known as the father of geology. He played a key role in establishing geology as a modern science. He attempted to formulate geological principles based on observation of the rocks.

Inge Lehmann (1888-1993) was a Danish seismologist and geophysicist who is known for her theory that the Earth consists of three shells: the mantle, outer core and inner core. She analyzed earthquake waves to discover that the earth's center is a solid core with diameter of more than 1,000 km.

Alfred Wegener (1880-1930) was a German meteorologist geologist, and geophysicist, best known as the author of the continental drift theory, a substantial basis of today's plate tectonics. He is known as pioneer of polar research and did four Greenland expeditions.

Since the Industrial Revolution in the 1800's, there are millions of factories operating all over the world. These manufacturing processes makes our daily needs like food, papers, clothes, everything. Since then, there was also an increase in environmental regulations or rules the factories have to follow in order to be given a license to operate.

Environmental geologists work to solve problems associated with pollution, waste management, urbanization, and natural hazards, such as flooding and erosion. They deal with issues of water, both surface waters and groundwater.

Others work with soil, managing this most important resource for the sustainability of Agriculture such as soil erosion and poor soil health. Another area of study relating to soil are natural hazards such as landslides.

Geoscientist has a unique role as steward of our water and soil resources. But it is equally our responsibility to guard the natural resources around us.

An Economic or Resource geologist is responsible for geological assessment like environmental impact. They analyzed geological information to identify sites that should be explored. They develop strategies, do resource modelling and make informed decisions about a certain project Generally this projects are to mine minerals and natural oils or gas. Their assessment must be accurate to avoid environmental issues like destruction of land or sea ecosystem, landslides etc.

There must be a good balance between conservation and usage. Without these resources, our lives will not be as easy. Just imagine life without phone, airplane and elevator. So these two geologists should work together for the benefit of both humankind and our environment.

Skills needed for this job are Data collection, Analytical, Fieldwork skills, Detail-oriented, both oral and written communication skills. Geologists are skilled navigators. Do you think this job is right for you?

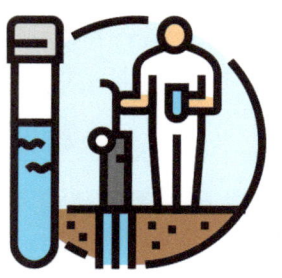

The Geographer

Geography is the study of the Earth and the forces that shape it. It specialised on Earth's landscapes, people, places and environment. There are two main branch of Geography: Physical geography and Human geography.

The five themes of geography are location, place, region, movement, and human- environment interaction. The five themes enable you to discuss and explain people, places, and environments of the past and present.

Physical Geographers study landscape, water, soil and climate. They study processes and patterns in the environment such as the atmosphere, hydrosphere, biosphere and geosphere.

Biosphere is made up of the part of the earth where life exists

Atmosphere is made of layers of gasses that envelops the Earth

Hydrosphere is the sum of all the Earth's water (on, under & above)

Geosphere is the earth itself and what it is made of; both surface & interior

Human Geographer studies the spatial relationships between human communities, cultures, economies and their interactions with the environment. They studies human behaviours like how people organise their communities, gather and distribute resources and adaptability to their environment. The following are some of the specialisations.

Economic Geography: the study of how wealth and resources are distributed in different regions. Geographers identify the physical factors that influence the economic development patterns.

Urban Geography is the study of cities. Geographers identify potential areas that are ideal for land development and growing infrastructure for the populations.

Political Geography is focusing on how the countries set their borders for management and control purposes. This includes national borders and local boundaries or central and local governments.

Medical Geography is the study how disease spreads through different parts of the world. Geographers studies how pandemics and other common diseases reach communities around the globe.

Military Geography is the study of the ideal conditions for distributing military facilities, troops and supplies. These research are based on terrains, transportation infrastructure and climate.

To learn geography it is important to learn directions. Geographers used geodetic coordinates in longitude and latitude for measuring and communicating positions. The invention of a Geodetic Coordinate System is generally attributed to Eratosthenes.

Eratosthenes was a Greek Polymath; a mathematician, geographer, astronomer among others. He is known as the father of Geography, famous for measuring the diameter of the earth.

Geographers spend most of their time gathering and analysing information. They collect data by interpreting satellite images, census reports and field observations. They gather samples and analyse in the lab. Geographers creates and updates maps of certain phenomena as well as making predictions about how the environment will develop over time.

Geographers must be skilled in GIS programming (Geodetic Information System), data management and data visualisation techniques. Communication skills is important as they often need to write reports and proposals. Critical thinking skills, analytical skills too, among others.

Are you an investigative person? This maybe a future career path for you. Think about it!

www.ingramcontent.com/pod-product-compliance
Lightning Source LLC
Chambersburg PA
CBHW040250220526
45473CB00001B/436